MICAH CAN DO DUUUVAL MATH

ADRIAN MCCLINTON

Mamba Math LLC

MICAH CAN DO DUUUVAL MATH

1 Addition

2 Subtraction

3 Multiplication

4 Division

5 Special

CONTENTS

Achieve your dreams

If you can imagine it, you can picture it.

If you can picture it, you can see it.

If you can see it, you can explain it.

If you can explain it, you can plan it.

If you can plan it, you can pursue it.

If you can pursue it, you can achieve it!

Achieve your dreams!

by: Adrian McClinton

The Beginning

Hi everyone! My name is Micah, and I am passionate about math. I enjoy creating math equations wherever I go. I have this super cool notebook and pencil that always tag along with me. I even gave them special names: Mrs. Notebook and Mr. Pencil.

Every morning when Micah wakes up, he starts his day by writing down all sorts of different equations that will challenge his mind.

One morning, Micah opened his sock drawer and saw two rows of five pairs of socks. He picked a pair for his feet and wondered, "How many socks are left in the drawer?" Can you help him find the correct answer?

My Socks

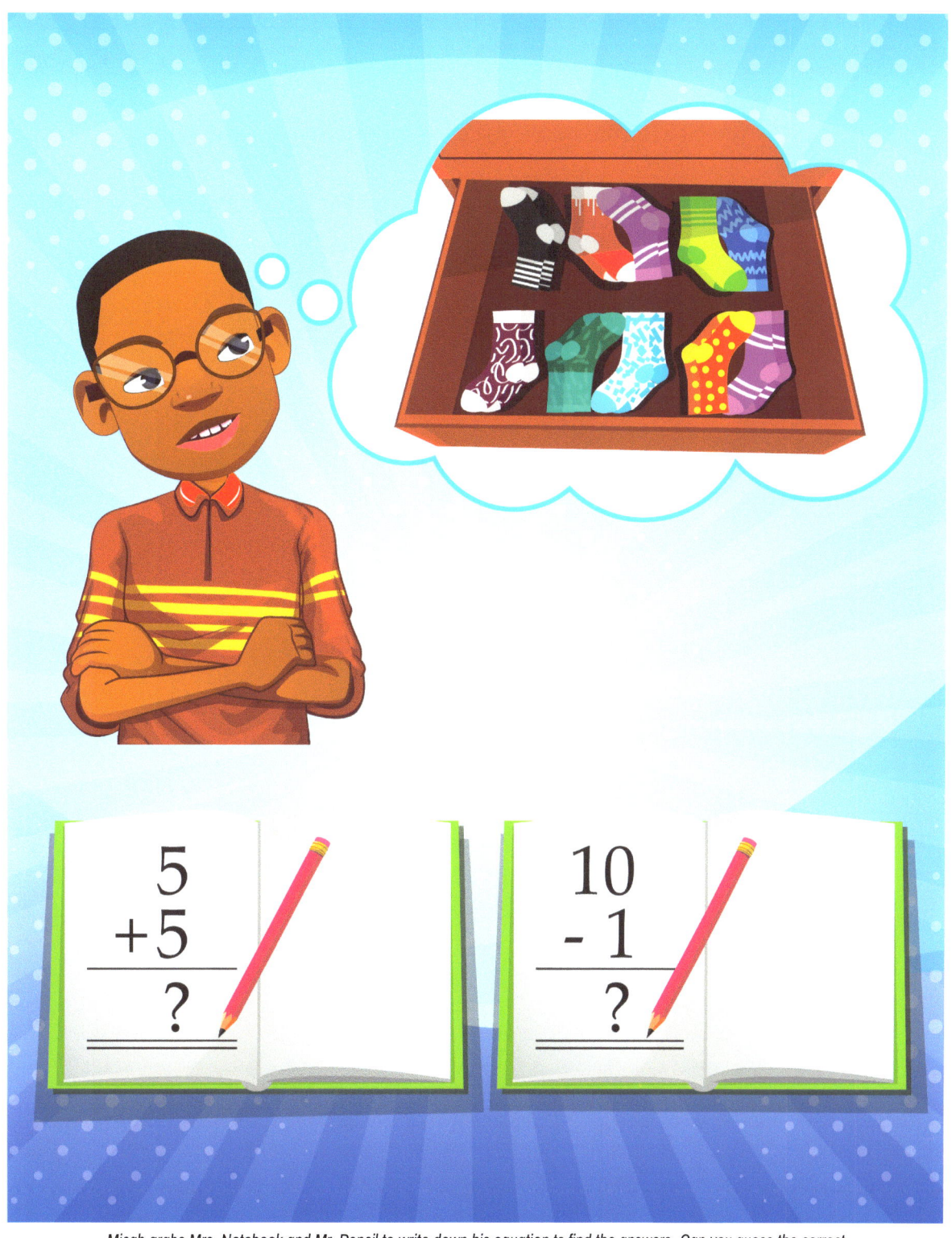

Micah grabs Mrs. Notebook and Mr. Pencil to write down his equation to find the answers. Can you guess the correct answers?

Teeth Cleaning Routine

As Micah put on his socks and slippers, he headed to the bathroom.

First, he gathered all the supplies required for cleaning his teeth. So, he grabbed a cup, mouth-wash, floss, toothpaste, and his toothbrush.

He had everything necessary to begin, but he wondered what the total sum of all the items, including the bathroom mirror he uses, would be.

Good Hygiene

Items	How Many
Cup	1
Floss	1
Toothbrush	1
Toothpaste	1
Mouthwash	1
Mirror	1
Total	?

Micah grabs Mrs. Notebook and Mr. Pencil to write down his equation to find the answers. Can you guess the correct answers?

Micah Wears Uniforms To School

"I love my school!" he replied. "One of my favorite things about it is that we all wear uniforms."

All the students dress alike, wearing the same colors. A shirt and a pair of pants make one complete uniform for Micah.

As Micah gets ready for school, he looks into his drawer and sees how many uniforms he has. He notices three rows of three shirts, all neatly folded up. Next to the shirts, he sees three rows of three pairs of pants.

Now, Micah wonders how many uniforms he has left for school if he takes one complete uniform from his dresser.

School Uniform

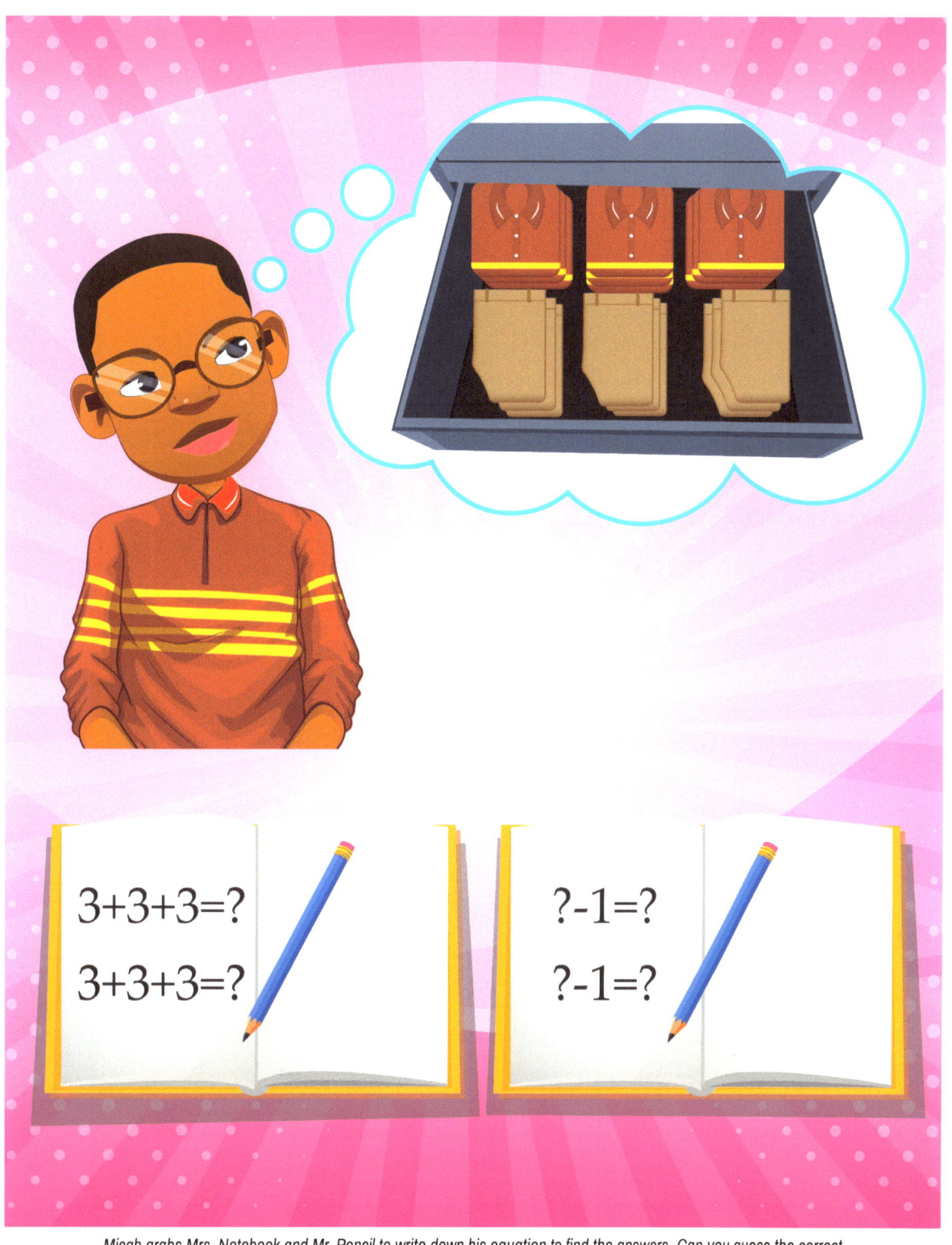

Micah grabs Mrs. Notebook and Mr. Pencil to write down his equation to find the answers. Can you guess the correct answers?

Breakfast

Micah loves eating breakfast. Once he is ready for school, he heads downstairs to watch his mom make breakfast.

Since he is the first one downstairs, she makes his food first.

He noticed that she had a case of a dozen eggs, and she removed two from the case to cook them. Along with the eggs, she also cooked some bacon, made toast, and poured him a glass of almond milk. As he enjoyed his meal, he wondered how many eggs were left in the carton.

Micah Breakfast

Micah grabs Mrs. Notebook and Mr. Pencil to write down his equation to find the answer. Can you guess the correct answer?

Micah's Brotherly Love

Micah's four brothers are all getting ready for school.
All of the brothers have their own unique personalities, which make them truly special individuals.

The oldest brother, James, is a responsible seventeen-year-old who loves to dance.
Right after him is Donovan, who is sixteen years old. He is talented and hardworking.

Next is AJ at fifteen, who possesses a great sense of humor and is also a track star.

And then there is Mason, the youngest of them all. He is just a playful seven-year-old with natural raw talent. As for Micah, he is nine years old, the middle child in this lively bunch, who loves math and fishing. What is the age difference between James and Micah?

Micah's Tribe Of Brothers

Micah grabs Mrs. Notebook and Mr. Pencil to write down his equation to find the answer. Can you guess the correct answer?

Moms School Adventure

Micah's mom takes everyone to their respective schools. Mason and Micah attend the same elementary school. It's five blocks away from their home.

AJ, however, attends middle school, which is located a bit further, around fifteen blocks away from their house.

As for James and Donovan, they both go to the same high school, which is thirty blocks away from their home.

Micah wonders, how many blocks does his mom drive in total to reach all three schools?

Micah's School

Micah grabs Mrs. Notebook and Mr. Pencil to write down his equation to find the correct answer. Can you guess the correct answer?

So Many Traffic lights

Wow, there are so many traffic lights! Micah, exclaims!

After school, my mom picks us all up. She always takes the same route to pick up everyone.

First, we go past three traffic lights before my mom turns left. Then, we go through four more traffic lights to reach AJ's school.

Next, we pass six traffic lights before she turns right. Finally, we travel through five more traffic lights until we reach James and Donovan's school.

Can you guess how many traffic lights we passed in total?

So Many Stop Lights

Micah grabs Mrs. Notebook and Mr. Pencil to write down his equation to find the answer. Can you guess the correct answer?

Review

Let's review:

1. How many socks are left in the drawer? 9

2. What is the sum of all the items together including the bathroom mirror Micah used? 6

3. How many uniforms does Micah have left in his dresser for school? 8

4. How many eggs were left over? 10

5. How many years is James older than Micah? 8

6. What is the total number of blocks for all three schools from their house? 50

7. How many traffic lights did they pass altogether? 18

Summary

How many socks are left in the drawer?

What is the sum of all the items together including the bathroom mirror Micah used?

How many uniforms does Micah have left in his dresser for school?

How many eggs were left over?

How many years is James older than Micah?

How many traffic lights did they pass altogether?

What is the total number of blocks for all three schools from their house?

Adrian McClinton, born on February 28, 1986, in Jackson, Mississippi. He overcame a challenging childhood, having grown up in a low-income neighborhood and being raised by a single parent. In sixth grade, he joined the Young People Project (YPP), a Math literacy program founded by Dr. Robert Moses, also known as Bob Moses, who is renowned for establishing the Algebra Project. Through YPP and the Algebra Project, Adrian and other young children created engaging math games to assist struggling students in grasping mathematical concepts in an enjoyable manner. This initiative also kept the young children off the streets and allowed them to make positive impacts in their communities.

Adrian's passion for education and teaching mathematics persisted throughout middle and high school. This was key in the decision to enlist in the United States Navy after graduating. His military service granted him extraordinary life experiences and the privilege to explore awe-inspiring destinations that most people can only dream of. The Navy played a pivotal role in shaping Adrian into a resilient leader, further enhancing his personal growth.

While maintaining active duty status in the US Navy, Adrian takes immense pride in being a devoted father to five sons, including Micah, his second youngest. When Adrian was relocated to Florida, Micah encountered difficulties with math during his initial year of school. Recognizing this as an opportunity to support his son, Adrian utilized his teaching expertise and incorporated math into their everyday lives. This marked the inception of Micah Can Do Duuuval Math.

Adrian's ultimate objective is to make math relatable to all children by integrating it into their daily routines. Through their family's unique journey, he aspires to positively impact the lives of millions of children worldwide. Adrian endeavors to ignite a passionate drive within those who perceive math as daunting, encouraging them to embrace math challenges with confidence rather than intimidation.